Course Overview

This course presents an overview of various modern electrical substation engineering aspects, which may be very challenging and require from utility company's engineers a very diverse knowledge and experience to address them.

The entire course consists of the following five sections:

- Overview of Typical Substation Switching Systems

- Air-Insulated (AIS) Versus Gas-Insulated (GIS) Substations

- Substation Safety and Fire Protection

- Substation Physical Security

- Substation Insulator's Performance Improvement

Each section covers a specific topic chosen by the author and includes a description of the problems associated with this topic as well as their possible resolution.

While this list of topics does not include all technical issues that substation engineer may encounter, it gives a good idea about their complexity and importance for effective operation of power systems.

This course is beneficial for any engineering, management, and educational professionals looking to expand their knowledge and experience in modern electrical substation engineering, most notably:

- Electrical engineers and designers

- Colleges and universities faculty

- Utility company project management personnel

- Construction professionals

To facilitate self-learning and improve retention of material, quizzes are provided in the end of each section. Quiz answers are included as well in the end of the course material.

Information about additional services offered by the author may be found on the website: www.substationguru.com

Section 1

Overview of Typical Substation Switching Systems

As the author indicated in [1], two of electrical substation main missions are:

- Connection of separate transmission and distribution lines into a system to increase the efficiency and reliability of power supply

- Sectionalizing of power system to increase its reliability and operational flexibility

To fulfill these missions, each substation should have a switching system which consists of lines connected together by one or another type of substation bus. Each line is usually protected by the circuit breaker. Examples of substation bus and circuit breakers are shown in Figs. 1 and 2, respectively.

Fig. 1. 138 kV Open-Air Bus on Post Insulators

Fig. 2. 26 kV Outdoor 2000 A Circuit Breaker

For a detailed description of circuit breakers and substation busses including their mission and types, you may see corresponding sections of [1].

It is important to note here that the same number of lines associated with the substation may be connected together into different types of switching systems, each of them having their own advantages and disadvantages.

Because the selection of a specific switching system represents one of fundamental substation engineering tasks, it seems useful to the author to review their most popular types. To do it, let's consider all kinds of these systems for the substation with four lines on a high-voltage side and compare their advantages and disadvantages.

Let's start this review with the simplest system called a "Single Straight Bus System" shown in the Fig. 3.

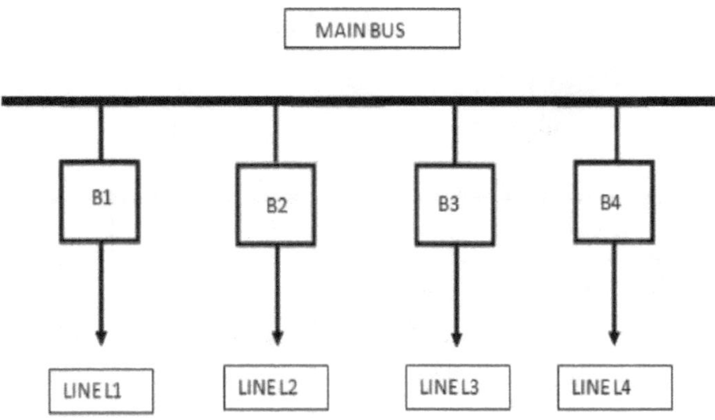

Fig. 3. Single Straight Bus System

This bus arrangement has the following advantages:

- The lowest cost out of all switching systems

- Simplicity

- It is easily expandable

- A small land area is required

However, there are the following disadvantages as well associated with this system:

- The lowest reliability out of all switching systems

- Bus of breaker fault causes a loss of an entire station

- A breaker removal for maintenance causes a loss of a corresponding line

Because of these deficiencies, single straight bus switching systems are rarely used for modern new substations. However, substations built more than 20 years ago still often have them. Even today, depending on how the whole system with interconnections between substations is arranged, a straight bus may be considered for a specific substation if reliability of the whole grid is acceptable.

The next substation bus arrangement that we want to consider is a "single sectionalized bus system," shown in Fig. 4, which has an additional sectionalizing breaker BS and provides a little bit more reliability and flexibility than just a straight bus.

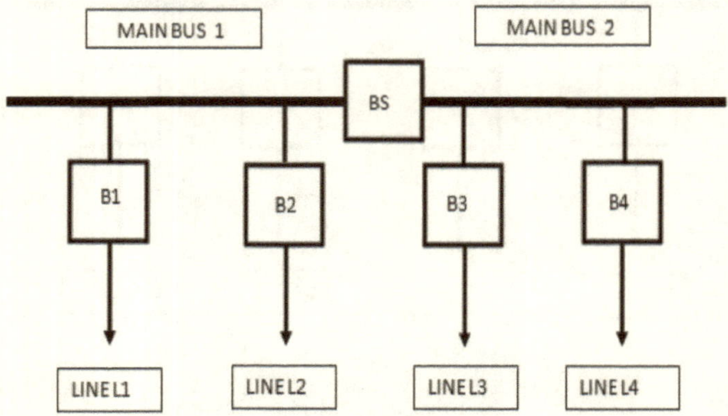

Fig. 4. Single Sectionalized Bus System

As discussed in [1], the main benefit of sectionalization of power systems is a reduction of customer outages associated with the fault on the line, transformer, etc., and this is exactly the case here. Comparing systems shown in Figs. 3 and 4, we can clearly see that if for a bus without a section breaker, a failure of any line breaker or the bus led to the outage of the whole station; for a bus with a section breaker, this type of fault will lead to the outage of only half of the station. The only situation when the whole station will be switched off is a failure of a bus section breaker.

Let's summarize pros and cons for a single sectionalized bus system:

- Advantages:

    - A low cost

    - Simplicity

    - It is easily expandable

    - A small land area is required

    - A higher reliability than a single bus system

- Disadvantages:

    - An additional breaker is required

– A sectionalizing breaker's fault causes a loss of an entire station

– A breaker removal for maintenance causes a loss of a
corresponding line

Because of a low reliability, all of the abovementioned switching systems lost their former popularity and are rarely used at modern substations. One of the more reliable and widely used bus arrangements is a "ring bus system," shown in Fig. 5, which has two breakers associated with each outgoing circuit. Fault on any circuit will lead to both associated breakers tripping and breaking a ring, which is acceptable.

If any of the breakers fails, the adjacent breakers will trip, and two circuits will be taken out of service, but not the whole station, as it was the case for a straight bus without a section breaker. If we need to take any breaker out of service for maintenance or repair, the adjacent circuits will stay energized, unlike with both single straight busses with or without a section breaker, where taking out the line breaker leads to the outage on the associated line as well.

Let's summarize pros and cons for a ring bus system:

- Advantages:

    – A flexible operation

    – A high reliability

    – A double feed for each circuit

    – Any breaker may be removed without affecting the service

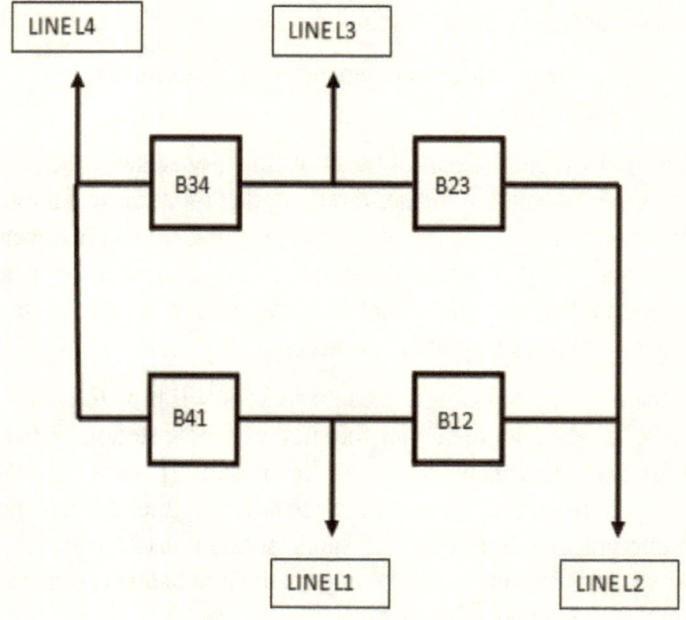

Fig. 5. Ring Bus System

- Disadvantages:

    – Breakers should have a high current rating based on a total load

    – Difficult to arrange

    – Any line fault trips two breakers

If a reliability analysis shows that a ring bus is not reliable enough, the next step in increasing redundancy of power supply is an application of a so-called "breaker-and-a-half system," shown in Fig. 6.

The strange name for this bus arrangement is based on the fact that each bay of this switching system (there are two bays shown in Fig. 6, but usually a substation switchyard consists of several bays) has three (3) breakers for two (2) circuits. So, if you want to find formally how many breakers are associated with each circuit, this number will be 3/2, or there are 1.5 breakers per circuit. Unlike the case with a ring bus system, where a failure of any breaker led to taking out of service two adjacent circuits, for a breaker-and-a-half arrangement, only a failure of a middle breaker in the bay will lead to losing two circuits connected on each side of this breaker. For a failure of a breaker connected to any of the main tie busses (left or right), only one circuit associated with this breaker will be switched off.

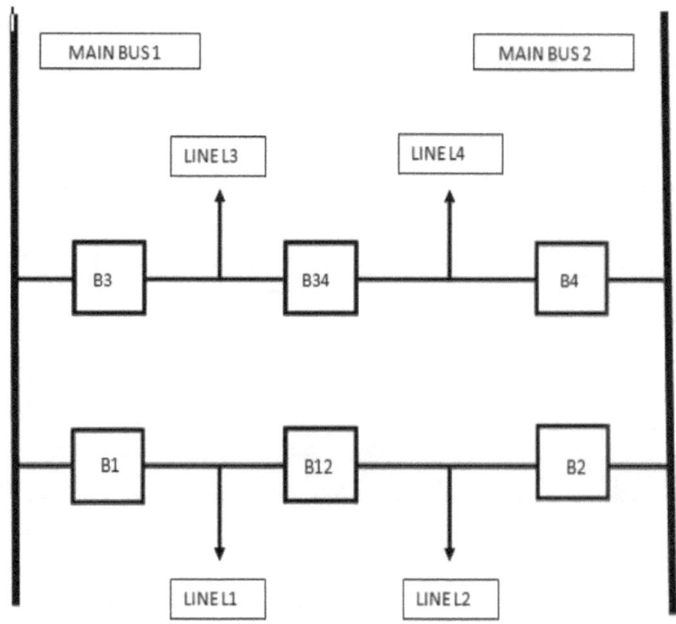

Fig. 6. Breaker-and-a-Half Switching System

Advantages and disadvantages of a breaker-and-a-half system may be summarized as follows:

- Advantages:
    - A flexible operation
    - A high reliability
    - A double feed for each circuit
    - Any breaker may be removed without affecting the service
    - A tie bus fault doesn't interrupt service to any circuit
    - All switching is done with circuit breakers
- Disadvantages:

– 1 ½ breakers are required per circuit

– A complicated relaying

– A middle breaker fault causes a loss of two adjacent circuits

Again, if a reliability analysis still requires a higher redundancy than a breaker-and-a-half arrangement provides, there is one more option left: it is the so-called a "double bus–double breaker system," shown in Fig. 7.

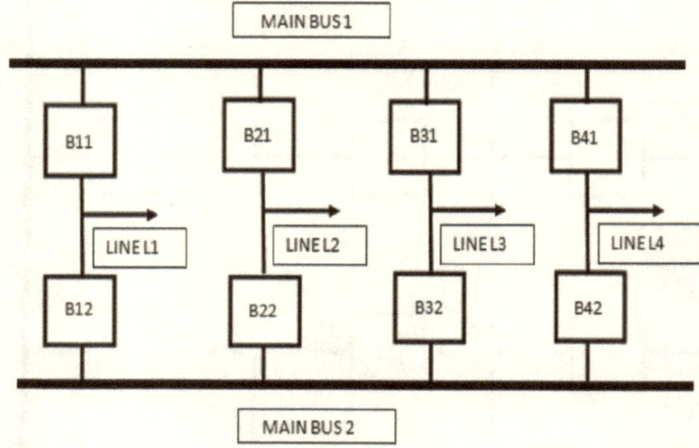

Fig. 7. Double Bus–Double Breaker System

As a step forward reliability-wise, for a double bus–double breaker system, a failure of any breaker will never lead to the loss of more than one circuit, while for a breaker-and-a-half a failure of a middle breaker resulted in two circuits being switched off.

Let's summarize pros and cons of a double bus–double breaker switching system.

• Advantages:

– A flexible operation

– A very high reliability

– A double feed for each circuit

– Any breaker may be removed without affecting the service

– A bus fault doesn't interrupt service to any circuit

– All switching is done with circuit breakers

– Only one circuit is lost for a breaker failure

• Disadvantages:

– A high cost.

– Two breakers are required for each circuit.

Finalizing our discussion about typical substation switching systems, I wanted to emphasize that a selection of any specific bus arrangement needs to be based on a financial analysis, because any increase in reliability will lead to extra costs, and this investment should be justified before the more reliable switching system is chosen. It may be noted that out of all abovementioned bus arrangements, the most popular one used in modern substations is "breaker-and-a-half," because it is providing high reliability at a reasonable cost.

Section 1 Quiz

1. True or False: The least expensive substation switching system is a single sectionalized bus system.

    a) True

    b) False

2. How many circuit breakers does the ring bus system for six lines have?

    a) 4

    b) 5

    c) 6

    d) 7

3. What is the most reliable substation switching system?

    a) Ring bus

    b) Breaker-and-a-half

    c) Double bus–double breaker

    d) Single straight bus

4. Which substation switching system is easily expandable?

   a) Single straight bus

   b) Single sectionalized bus system

   c) Both of them

   d) None of them

5. If an engineer has to choose a substation switching system fitting into the smallest land area, which one should he select?

   a) Ring bus

   b) Breaker-and-a-half

   c) Double bus–double breaker

   d) Single straight bus

Section 2

Air-Insulated (AIS) Versus Gas-Insulated (GIS) Substations

After substation engineer chose one or the other switching system for a new substation, he needs to design its key plan, which should show location of all components of a substation and their interconnections, as well as steel structures, a control house, fire walls, driveways, a fence, and a property line.

Working on a substation key plan, engineers are frequently running into the problem of fitting all substation components into an available land plot. The first choice here should be the so-called air-insulated substation (AIS) where all connections between substation components (transformers, breakers, disconnecting switches, etc.) are made in an open air because it is the most economical solution. However, this type of substation requires for its implementation an ample land space which is not always available.

If this is the case, a gas-insulated substation (GIS) may be considered instead of AIS. GIS is a high-voltage substation in which all the components except power transformers (bus, circuit breakers, disconnecting switches, etc.) are contained in a sealed environment filled with sulfur hexafluoride gas (SF6) as the insulating medium. Because SF6 is a much better insulator than air, all minimum phase-to-ground and phase-to-phase clearances required for a specific voltage level may be significantly reduced, and as a result, we may develop a substation of required configuration on approximately two times smaller land plot than the one needed for AIS.

For better understanding, let's compare a 230 kV breaker position in air-insulated station shown in Fig. 8 with several 420 kV breakers in a gas-insulated station shown in Fig. 9.

Fig. 8. 230 kV Circuit Breaker in AIS

Fig. 9. 420 kV Circuit Breakers in GIS[1]

At air-insulated station (Fig.8), only circuit breaker's interrupter is enclosed in a steel tank filled with SF6 gas, while disconnecting switches and bus are open air.

At gas-insulated station (Fig. 9), circuit breakers, disconnecting switches, and bus are located inside steel enclosures filled with SF6. Gas-to-air bushings are used to connect GIS to the rest of substation.

GIS has the following advantages compared with AIS:

- Higher reliability
- Protection from weather, animal contact, etc.
- Need for a small land area
- Low maintenance (especially for SF6 bus)

Disadvantages of GIS versus AIS:

- High cost
- Inability to detect a gas leak (SF6 bus)
- Difficult troubleshooting
- Long repair time
- Possible maintenance personnel resistance to learning new equipment

Section 2 Quiz

1. True or False: Compared with AIS, GIS requires the smaller land plot for substation of the same configuration and is less expensive.

    a) True

    b) False

2. Which AIS components have open-air connections?

    a) Circuit breakers

    b) Disconnecting switches

    c) Transformers

    d) All of them

3. Which GIS components are contained in a sealed environment?

a) Circuit breakers

b) Disconnecting switches

c) Bus

d) All of them

4. What are AIS advantages over GIS?

a) Higher reliability

b) Protection from weather, animal contact, etc.

c) Need for a small land area

d) None of them

5. When do we use GIS instead of AIS?

a) When we need to reduce substation footprint

b) When we want higher reliability of power supply

c) Both a and b

d) Neither a nor b

---

[1] By I, Dingy, CC BY-SA 3.0, https://commons.wikimedia.org/w/index.php?curid=2373691

Section 3

Substation Safety and Fire Protection

It is fair to say that safety is always a # 1 priority in substation design, engineering, operation, and maintenance. Unlike it was the case with reliability where a higher reliability required a larger investment, we can't put a price tag on safety, because there is no such thing like working conditions being more or less safe – it is should be 100% safe to work at the substation or just to visit it. There are numerous laws, rules, codes, etc., governing safety requirements, one of the most important being "IEEE Standard C2-2017" [2]. Main mission of all these regulations is safeguarding of persons from hazards, arising from the installation, maintenance, or operation of substation equipment. Safety standards contain requirements for:

- Enclosure of electrical equipment

- Rooms and spaces

- Illumination

- Floors, floor openings, passageways, stairs

- Exits

- Installation of equipment:

    – Protective grounding

    – Guarding live parts

    – Working space above electrical equipment

- Specific rules for installation of all typical substation equipment

All these measures are based on common sense and the goal to provide a safe environment for substation personnel. The following requirements may be mentioned as an example:

1. Enough clearance from energized parts should be provided to avoid accidental contact with them. If that can't be met, live parts should be

guarded or enclosed.

2. A minimum height from the ground to any ungrounded part of electrical installation should be 8'–6", so a person staying on the ground can't touch a substation element or its part which may become energized accidentally. For example, a bottom of post insulator supporting energized bus does not normally has any potential. However, if bus flashover to the ground over insulator occurs, touching a bottom of insulator may become unsafe. That's why 8'–6" distance from the bottom of insulator to the ground should be provided.

It is important to mention here that if there is a platform above ground for somebody to stand on it, 8'–6" to the bottom of insulator should be measured from the top of the platform rather from the ground. Let's take a look at the bottom of a steel column shown in Fig. 10. Let's assume that there is a post insulator installed on the top of this column. In this case, a safety clearance of 8'–6" to the bottom of the insulator should be measured from the top of the concrete footing or even from the steel plate of the column but not from the ground, because hypothetically somebody can stand on the top of the footing.

Fig. 10. Installation of a Steel Column on Concrete Foundation

3. There should be sufficient illumination for personnel to clearly see their surroundings and perform any work safely. Required illumination levels are specified in [2].

4. All passageways and stairs should be wide enough for personnel to navigate them safely, adequate railing should be provided, and floor openings should have guard rails.

5. Exits should be clearly marked, and evacuation routs should be free from obstructions.

Depending on a function of the building (e.g., control house), it may require several exits to avoid personnel being trapped during equipment fault, fire, etc.

6. All substation metallic structures, fences, and equipment tanks should be connected to a station ground grid, which should be designed to ensure that step and touch potential values are lower than the ones stipulated in the applicable standards.

For more information about substation grounding, please see IEEE Std. 80-2013 [3]. Another important topic related to substation safety is fire protection whose mission is to protect substation personnel, equipment, and buildings from fire and prevent fire from spreading. As it was with safety, there are numerous guides and standards for substation fire protection, most notably IEEE Std. 979-2012 [4].

The following means of fire protection are used:

- Separation of equipment

- Installation of protective fire walls

- Deluge systems

- Application of fire retardant materials for substation buildings

- Fire alarm system

- Installation of oil retention pits for oil containing equipment

- Appropriate substation design to prevent fire from spreading from one part of the station to another, for example, avoidance of an uneven substation surface profile.

Out of this list I wanted to discuss briefly the first item, i.e., separation of equipment, because of frequent confusion on how to measure this separation. In IEEE 979-2012, there are numerous tables telling you how far apart pieces of oil containing equipment (mostly transformers) should be kept to avoid a need for firewall between them. This minimum separation distance depends on amount of oil in equipment. For example, you can find that for transformers with 10000 and more gallons of oil inside, a fire wall between them is not required if they are separated by at least 25 feet. Question is how to measure this separation distance. Let's take a look at two transformers shown in Fig. 11 There is an oil containment under each of them built in accordance with IEEE 980-2013 [5]. This containment is sized to accept all the oil that transformer has in case transformer is leaking oil or went on fire and now oil containment is filled with a burning oil. The separation distance between transformers should be measured not between transformer tanks ("skin to skin"), but between a pool of fire created by burning transformer oil in Transformer #1 containment and Transformer's #2 tank, and vice versa (see Fig. 11).

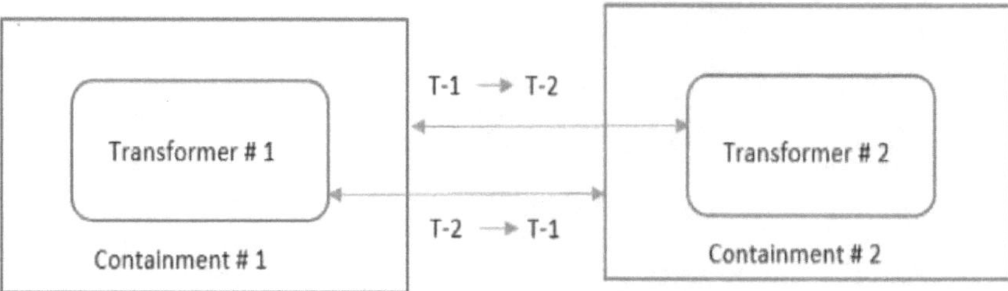

Fig. 11. Measuring the Separation Between Oil Containing Equipment

As we can see, a separation found using a pool of fire approach is lower than a distance between transformer tanks, which may result in a need for firewall where we did not need it before while measuring a separation "skin to skin."

If a separation distance specified by the standard can't be provided, a fire wall made out of a fire resistant material (usually having 2-hour rate of a fire resistance) should be built between transformers. The example of a fire wall protecting different substation components from the fire on each of them is shown in Fig. 12.

Fig. 12. 138 kV Cable Oil Pump House with a Fire Wall

Section 3 Quiz

1. True or False: You can deviate from safety standards in order to finish substation project on time and under budget:.

    a) True

    b) False

2. What safety measures will protect personnel from accidental contact with energized parts?

    a) Having live parts enclosed

    b) Providing sufficient clearance from the live parts

    c) Having live parts guarded

    d) All of them

3. What are the means of substation fire protection?

    a) Separation of equipment

    b) Installation of protective fire walls

c) Deluge systems

d) All of them

4. What separation distance between transformers containing more than 10000 gallons of oil will allow not to have a fire wall between them?

a) 25 feet

b) 30 feet

c) Both a and b

d) Neither a nor b

5. What rate of fire resistance should material used for a fire wall have?

a) 2-hour rate

b) 1-hour rate

c) 0.5-hour rate

d) All of them

Section 4

Substation Physical Security

This issue became so important lately that there is an IEEE Std. P1402 (Draft) [6], which is devoted to substation physical security. Here I would like to mention two most common problems whose solutions may greatly affect substation detail design.

The first problem is a copper theft which became a worldwide problem in the last 20 years. To understand what thieves do, let's take another look at a steel column shown in Fig. 10, this time concentrating on a typical ground cable connection to it. To make it easier for you to follow a description of the problem, I repeatedly showed this column in Fig. 13.

Thieves are getting into the station either by climbing the fence or by taking the fence down by driving a truck thru it. After that, they cut a copper ground cable that you see in Fig. 13 at the column, pull cable out of the ground as far as they can (usually, around 5 feet), and cut it on the ground grid side. Doing it multiple times during one raid on the station, thieves can cut hundreds of cables, leaving substation equipment and structures ungrounded. As a result of that, people inside and outside of substation are not protected anymore from excessive step and touch potential and may be electrocuted during ground fault.

Fig. 13. Ground Cable Connection to a Steel Column

Another problem is associated with possible damage to equipment and structures. For example, when short circuit to ground happens, a fault current will return to the source not thru a low-resistance ground grid but thru the ground itself, foundation rebars, etc., which may damage foundations and create dangerous overvoltages damaging the equipment. For more information about substation grounding, please see IEEE Std. 80-2013 [3].

There are several ways to fight this problem. One of them is to hide a copper cable out of the plain view, locating it underground. To accomplish that, we can use a steel detail which is welded above ground to a base plate of the steel column shown in Fig. 13 and is cad welded underground to a copper ground cable. As a result of doing that, we'll have steel columns without visible connections to the ground grid, as shown in Fig. 14.

Fig. 14. Steel Structure with Anti-theft Detail Installed

Another way to fight a copper theft is to use a different material (with no scrap value) to create a ground grid and connect above ground structures to the grid, and let prospective thieves know that there is nothing inside substation that may interest them. The example of such material is copperweld which is a copper-coated steel wire. Substation which is using this approach is shown in Fig. 15.

Fig. 15. Substation with a Copperweld Ground Grid and Anti-theft

Another substation security issue that I wanted to mention is a protection of substation property from intruders by designing a perimeter fence appropriately. Usually, a substation is surrounded with a 7-ft high-chain link fence, specified in [7] with additional 1-ft-high section on the top made out of a barb or razor wire. However, very often this design is not accepted by local municipalities because of esthetical concerns. If this is the case, the upper 1-ft section of the fence may be made out of the same material as the rest of the fence and installed under approximately 45-degree angle in direction away from the substation property (see Fig. 16).

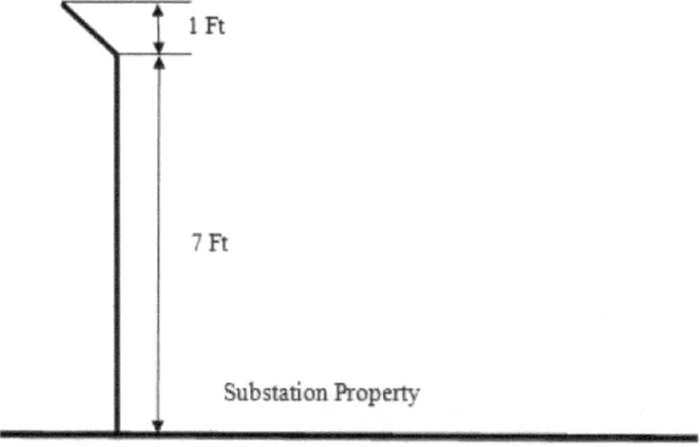

Fig 16. Substation Perimeter Fence with Improved Security

Section 4 Quiz

1. Substation physical security is covered in the following IEEE Standards:

    a) Std. C2-2017

    b) Std. 979-2012

    c) Std. 80-2013

    d) None of them

2. The theft of copper ground cables may result in the following in case of a ground fault:

a) Dangerous step and touch potentials

b) Possible damage to equipment and structures

c) Both a) and b)

d) Neither a) nor b)

3. The ways to fight copper theft are:

a) Hiding ground cables out of the plain view

b) Using copperweld for ground cables instead of copper

c) Installing anti-theft posters

d) All of them

4. The standard height of a substation chain link fence is:

a) 7 ft

b) 8 ft

c) 7.5 ft

d) 9 ft

5. To protect substation from intruders, the upper 1-ft section of substation fence may be:

a) Made out of a barb wire

b) Installed under 45-degree angle in direction away from substation property

c) Made out of a razor wire

d) All of them

Section 5

Substation Insulator's Performance Improvement

The last substation engineering challenge that I arbitrarily chose for consideration is insulator's performance improvement. As we know, insulation needs to be provided between energized parts and the ground as well as between phases of the same circuit or different circuits. One of the substation components providing a required insulation between energized conductors and grounded structures are substation insulators, whose main types are as follows:

- Station post insulators

- Equipment (transformer, breaker) bushings

- Equipment (CCVT, Surge arrester, etc.) insulating columns

For description of different kinds of substation equipment, see [1].

Insulator materials are either porcelain or polymer. For each insulator, there are numerous electrical and mechanical properties specified, which are used by substation designers and engineers in their calculations.

Typical porcelain station post insulator is shown in Fig. 17.

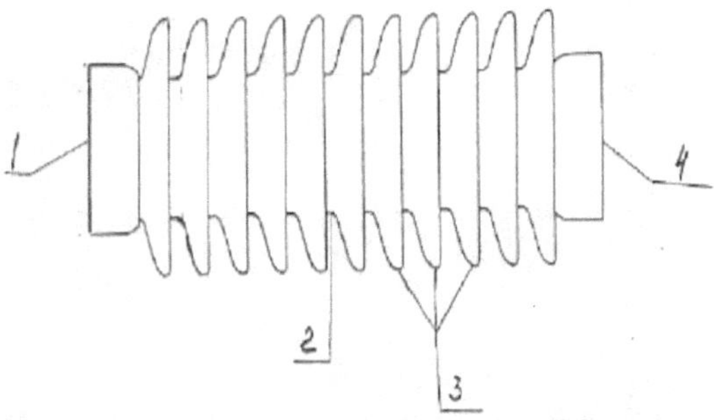

1 – Cap;   2 – Porcelain body;   3 – Skirts (Pettycoats);   4 – Base

Fig. 17. Station Post Insulator

Substation design is based on the assumption that substation insulators will retain their insulating abilities, but in reality, there are numerous factors that may affect insulator performance and therefore make them unsuitable for an intended application. Let's discuss the most common problems related to substation insulators and possible ways to resolve them.

One of these problems is icing, which represents a buildup of ice along the surface of insulators which is bridging the petticoats. This results in flashovers along the surface of insulators and equipment damage. Icing may be prevented by the following measures:

- Insulator surface coating (usually with silicon), which prevents ice buildup.

- Application of "high creep" insulators, with longer than standard skirts, which prevents their bridging by ice.

- Installation of creepage extenders, which are mostly used on existing insulators after it was discovered that they are susceptible to icing. Post insulators with extenders are shown in Fig. 18.

- Application of RGI (resistance graded insulators), which have a conductive coating, providing a path for leakage currents warming a surface of insulator body and by that preventing icing.

Fig. 18. 138 kV Post Insulators with Creepage Extenders

The next problem with insulators frequently encountered by substation maintenance personnel is pollution, which represents a deposit of pollutants (salt, dust, ashes, etc.) on the surface of insulators, resulting in flashovers, tracking along the surface of insulator and ultimately leading to insulator and equipment damage. The main remedy for pollution naturally is cleaning and washing of insulators, which needs to be done periodically as a part of substation maintenance.

Another serious problem associated with insulator maintenance is prevention of animal contacts to energized substation busses and bare cables. The most common animal intruders are squirrels and raccoons, who for some reason are often claiming on substation structures and touching conductors while bypassing the supporting insulators.

Usually this problem is associated with distribution voltages – up to 26 kV.

The consequences of that encounters are flashovers leading to equipment damage. Because this problem is so widespread, there is an IEEE Std. 1264-2015 [8] fully devoted to animal deterrents for substations.

The following measures may be used to prevent animal contacts:

- Installation of rubber boots, sleeves, etc., over bus connectors

- Insulation of bare conductors by using insulating tape, conduits

- Application of insulating paint over metal surfaces: transformer tank, etc.

Example of implementation of these measures is shown in Fig. 19.

Fig. 19. 13/4 kV Transformer 13 kV Bushings and surge arresters with rubber boots

Section 5 Quiz

1. The main types of substation insulators are:

    a) Station post insulators

    b) Equipment bushings

    c) Equipment insulating columns

    d) All of them

2. The common problems associated with substation insulators are:

    a) Icing

    b) Pollution

    c) Animal contact

    d) All of them

3. Substation insulator icing may be prevented by:

    a) Coating of insulator surface

    b) Application of high creep insulators

    c) Installation of creepage extenders

    d) All of them

4. The IEEE Standard covering animal deterrents for substations is:

    a) Std. C2-2017

    b) Std. 979-2012

    c) Std. 1264-2015

    d) None of them

5. The measures to prevent animal contacts with energized parts are:

    a) Installation of rubber boots over bus connectors

    b) Insulation of bare conductors

    c) Application of insulating paint over metal surfaces

    d) All of them

Conclusion

This course provided an overview of modern substation engineering aspects, concentrating on their detail description and possible resolution options to enable you to:

- Describe typical substation switching systems, their advantages, and disadvantages
- Compare air-insulated (AIS) and gas-insulated (GIS) substations
- Clearly understand importance of safety in substation design and engineering
- List means of substation fire protection
- Describe substation physical security issues and their resolution options
- List common problems associated with substation insulators
- Describe how to improve substation insulator's performance

Quiz Answers

Section 1 Quiz Answers

1. b) False
2. c) 6
3. c) Double bus–double breaker
4. c) Both of them
5. d) Single straight bus

Section 2 Quiz Answers

1. b) False
2. d) All of them
3. d) All of them
4. d) None of them
5. c) Both a and b

Section 3 Quiz Answers

1. b) False
2. d) All of them
3. d) All of them
4. c) Both a and b
5. a) 2-hour rate

Section 4 Quiz Answers

1. d) None of them

2. c) Both a and b

3. d) All of them

4. b) 8 ft

5. d) All of them

Section 5 Quiz Answers

1. d) All of them

2. d) All of them

3. d) All of them

4. c) Std. 1264-2015

5. d) All of them

www.ingramcontent.com/pod-product-compliance
Lightning Source LLC
Chambersburg PA
CBHW051420130726
47989CB00007B/3004